The World of Silk: Unraveling the Secrets of Silk Farming

Table of contents

1. Description

"The World of Silk: Unraveling the Secrets of Silk Farming" takes readers on a captivating journey into the intricate realm of silk production. From the historical origins of silk to the modern techniques of silk farming, this book unveils the fascinating secrets behind one of the world's most luxurious fabrics. Delving into the life cycle of silkworms, the meticulous process of sericulture, and the global impact of silk trade, the narrative weaves together science, culture, and history to provide a comprehensive understanding of the intricate world behind the silky threads that have woven their way through civilizations for centuries. Whether you're a textile enthusiast or curious explorer, this book promises to unravel the rich tapestry of silk's story.

2. The history of silk farming

The history of silk farming is woven into the fabric of human civilization, originating in ancient China and spreading its allure across cultures and continents. The journey of sericulture and silk production unfolds through millennia, marking a profound impact on trade, cultural exchange, and economic development. Here's a detailed exploration of the history of silk farming:

Ancient Origins in China:

1. Legend of Silkworm Domestication:
   -The history of silk farming is often intertwined with the legendary tale of Empress Leizu in ancient China, around 2700 BCE. According to legend, she discovered silkworms and the process of silk production when a cocoon dropped into her hot tea.

2. Silk Production Secrets Guarded:
   -The Chinese guarded the secrets of silk production for centuries, creating a lucrative monopoly on silk trade. Exporting silkworms or silk-producing knowledge was punishable by death, contributing to the mystique surrounding silk.

Silk's Journey along the Silk Road:

3. Silk Road Trade Routes:
   -The famous Silk Road, established during the Han Dynasty (207 BCE–220 CE), became the

conduit for silk trade and cultural exchange between China and the Mediterranean.

4. Silk Spreads Westward:
   -Silk's introduction to regions like Persia, India, and the Roman Empire transformed trade dynamics. Silk became a symbol of luxury and was highly sought after by royalty and the elite.

Silk in Byzantium and the Middle Ages:

5. Byzantine Silk Production:
   -Byzantium (Eastern Roman Empire) became a major silk-producing center, with silk workshops established in Constantinople. The Byzantines guarded their silk secrets with strict regulations.

6. Silk in Islamic Civilization:
   -Islamic civilization played a pivotal role in silk cultivation and production, with centers in Persia and later in Spain. The exchange of silk-making techniques continued during this period.

Silk Reaches Europe:

7. Introduction to Europe:
   -The Silk Road and trade connections facilitated the spread of silk to medieval Europe. Initially, silk was a luxury item, worn by the aristocracy and used in ecclesiastical garments.

8. Silk Guilds and Renaissance:

-The Renaissance era witnessed the establishment of silk guilds in Europe, promoting the growth of sericulture and silk weaving. Italy, particularly Florence and Venice, became renowned for silk craftsmanship.

Silk in Southeast Asia:

9. Spread to Southeast Asia:
   -The influence of silk spread to Southeast Asia, where countries like Thailand and Cambodia embraced sericulture. Silk became woven into the cultural fabric of these regions.

Modern Silk Farming:

10. 19th Century and Industrialization:
   -The 19th century saw advancements in silk production with the introduction of mechanized processes and industrialization. Silk mills emerged, transforming sericulture into a more streamlined industry.

11. Global Silk Production Centers:
   -China remains the leading producer of silk globally, followed by countries like India, Brazil, and Vietnam. Each region contributes its unique varieties and techniques to the global silk market.

Challenges and Innovations:

12. Challenges in Modern Times:

-Modern silk farming faces challenges such as competition from synthetic fibers, environmental concerns, and ethical considerations.

13. Innovations in Sericulture:
    -Ongoing innovations include genetic engineering for disease-resistant silkworms, sustainable sericulture practices, and the development of specialty silks.

The history of silk farming is a testament to the transformative power of human ingenuity, trade, and cultural exchange. From its mythical origins in ancient China to its global presence today, silk remains a symbol of elegance, luxury, and the intricate interplay between nature and human craftsmanship.

3. Silkworm Life Cycle

The silkworm life cycle is a fascinating and intricate process that unfolds through four distinct stages: egg, larva, pupa, and adult moth.

1. Egg Stage:
   -The life cycle begins with the laying of tiny eggs by adult silk moths.
   -These eggs are typically laid on specially designed paper or leaves in a controlled environment.

2. Larva Stage:
   -Upon hatching, the silkworm emerges as a larva, commonly known as a caterpillar.
   -The larva feeds voraciously on the leaves of mulberry trees, the primary food source for silkworms.
   -During this stage, the silkworm undergoes multiple molting phases, shedding its outer skin to accommodate its growing body.

3. Pupa Stage:
   -Once the larva reaches a mature stage, it spins a protective cocoon around itself using silk threads produced by silk glands.
   -The silk is extruded through spinnerets located near the mouth of the silkworm.
   -The spinning process takes several days, during which the silkworm forms a tightly woven cocoon to shield itself during the pupal stage.

4. Adult Moth Stage:

   -The pupa undergoes metamorphosis inside the cocoon, transforming into an adult silk moth.

   -The adult moth secretes an enzyme to soften a portion of the cocoon, allowing it to emerge.

   -Once emerged, the adult moth seeks a mate for reproduction.

   -After mating, the female silk moth lays eggs, thus completing the life cycle and initiating a new generation.

Silkworms are particularly sensitive to environmental conditions, and the success of silk farming depends on factors such as temperature, humidity, and the quality of mulberry leaves provided for feeding. This intricate life cycle has been meticulously managed by silk farmers for centuries, contributing to the production of the highly prized silk fibers used in textiles worldwide.

4. Mulberry Cultivation

Mulberry cultivation is a crucial aspect of silk farming, as the leaves of the mulberry tree serve as the primary and essential food source for silkworms. The success of sericulture, or silk production, relies heavily on the careful cultivation and management of mulberry orchards. Here is a detailed overview:

1. Selection of Mulberry Varieties:
   -Different varieties of mulberry trees are cultivated, with the most common being Morus alba (white mulberry) and Morus indica (Indian mulberry).
   -Varieties are chosen based on climate, soil conditions, and the specific requirements of the silkworm species being raised.

2. Soil Preparation:
   - Mulberry trees thrive in well-drained soil with good fertility.
   - Prior to planting, the soil is prepared by plowing and incorporating organic matter to ensure optimal nutrient levels.

3. Planting Mulberry Trees:
   - Mulberry trees are typically propagated through cuttings or seeds.
   - They are planted in rows with adequate spacing to allow for proper growth and ease of management.

4. Pruning and Maintenance:
   - Regular pruning is essential to encourage lateral branching, making it easier for farmers to access leaves for harvesting.
   - Mulberry trees require consistent care, including watering, weeding, and protection from pests to ensure a healthy crop.

5. Harvesting Mulberry Leaves:
   - Silkworms prefer tender leaves, so harvesting is often done by hand to ensure the quality of the leaves.
   - Leaves are harvested periodically, providing a continuous supply of fresh foliage for the silkworms throughout their feeding stage.

6. Leaf Quality and Nutrition:
   - The nutritional content of mulberry leaves directly impacts the health and productivity of silkworms.
   - Farmers monitor factors such as leaf size, texture, and color to gauge the suitability of the leaves for silkworm consumption.

7. Seasonal Considerations:
   - Mulberry cultivation is influenced by seasonal changes, with optimal growth occurring during specific periods.
   - Farmers plan their planting and harvesting schedules to align with the life cycle of silkworms.

Mulberry cultivation is an art in itself, requiring expertise and precision. The health and vigor of mulberry trees directly translate into the quality of silk produced, making the careful tending of mulberry orchards a cornerstone of successful silk farming practices.

5. Silkworm Rearing

Silkworm rearing is a meticulous process integral to silk farming, involving the nurturing of silkworms through their various life stages. The success of silk production hinges on the careful management of silkworms, from their initial hatching to the spinning of silk cocoons. Here's a detailed overview of the silkworm rearing process:

1. Egg Hatching:
   - Silkworm rearing begins with the careful incubation of eggs laid by adult silk moths.
   - Eggs are kept in a controlled environment with optimal temperature and humidity until they hatch into larvae.

2. Larval Feeding Stage:
   - Once hatched, silkworms enter the larval stage or caterpillar phase.
   - Silkworms are fed with fresh mulberry leaves, meticulously chosen for their nutritional quality.
   - The feeding stage is crucial for the silkworms' growth and development, involving multiple molting phases.

3. Molting:
   - Silkworms undergo molting, shedding their outer skin to accommodate their growing bodies.
   - Molting occurs several times during the larval stage, and each phase is marked by increased feeding activity.

4. Cocoon Spinning:

- As the silkworms reach maturity, they enter the spinning phase where they construct protective cocoons.
- Silkworms produce silk threads through specialized spinnerets near their mouths, forming a tightly woven cocoon around themselves.
- The spinning process lasts for several days and results in the creation of a silk-encased pupa.

5. Pupal Stage:

- The pupa undergoes metamorphosis within the cocoon, transforming into an adult silk moth.
- The pupal stage is a critical period, and the quality of the cocoon influences the quality of the silk fibers.

6. Cocoon Harvesting:

- Once the spinning process is complete, the cocoons are carefully harvested.
- The harvested cocoons are subjected to a boiling process to soften the silk proteins and facilitate unraveling.

7. Silk Extraction and Processing:

- After boiling, the silk threads are carefully unraveled from the cocoon.
- The extracted silk is then processed and woven into various textiles, marking the culmination of the silkworm rearing process.

Silkworm rearing demands a keen understanding of the silkworm life cycle and requires meticulous attention to environmental conditions. Factors such as temperature, humidity, and the quality of mulberry leaves significantly impact the success of silkworm rearing. Through centuries of practice, silk farmers have honed their skills, contributing to the production of one of the world's most coveted and luxurious textiles.

6. Cocoon Formation

Cocoon formation is a pivotal stage in the life cycle of silkworms, playing a central role in the process of silk farming. This intricate process involves the spinning of silk threads by mature silkworms to create a protective cocoon. Here's a detailed exploration of cocoon formation in silk farming:

1. Preparation for Spinning:
   - As silkworms approach maturity, they exhibit specific behaviors indicating their readiness to spin cocoons.
   - Silkworms are often observed becoming more restless and cease feeding as they prepare for the cocooning phase.

2. Spinneret Secretion:
   - Silkworms possess specialized glands called spinnerets, located near their mouths.
   - These spinnerets secrete a fluid containing silk proteins, which solidify upon contact with the air, forming silk threads.

3. Silk Thread Formation:
   - Silkworms carefully position themselves in a specific pattern while extruding silk threads in a figure-eight motion.
   - The silk threads rapidly solidify, creating a framework that eventually forms the basis of the cocoon.

4. Cocoon Shape and Structure:
   - The silkworm's spinning process results in the creation of a three-dimensional, oval-shaped cocoon.
   - The cocoon is intricately woven, with multiple layers of silk threads providing strength and protection to the pupa inside.

5. Cocoon Attachment:
   - Silkworms anchor their cocoons to a substrate, such as twigs or straw, to ensure stability during the pupal stage.
   - The attachment is often made at both ends of the cocoon, forming a secure and suspended structure.

6. Cocoon Color and Texture:
   - The color and texture of the cocoon vary depending on the silkworm species and environmental conditions.
   - Factors such as temperature and humidity influence the silk's characteristics, contributing to the diversity of silk textures in the final product.

7. Silkworm Metamorphosis:
   - Inside the cocoon, the pupa undergoes metamorphosis, transforming into an adult silk moth.
   - The silk moth secretes an enzyme to soften a portion of the cocoon, allowing it to emerge after completing the transformation.

8. Cocoon Harvesting:
   - The harvested cocoons are carefully collected, and the pupa inside is often sacrificed during the silk extraction process.
   - In some cases, a portion of the cocoons may be left intact to allow the completion of the metamorphosis and the emergence of adult moths.

Cocoon formation showcases the remarkable silk-producing capabilities of silkworms and underscores the delicate balance maintained by silk farmers to nurture healthy silkworms and ensure the production of high-quality silk fibers.

7. Harvesting and Processing

Harvesting and processing mark the culmination of silk farming, transforming meticulously cultivated cocoons into the luxurious silk fibers that have captivated cultures for centuries. This intricate stage involves careful harvesting, cocoon preparation, and silk extraction. Here's a detailed exploration of the harvesting and processing steps in silk farming:

1. Cocoon Harvesting:
   - Harvesting begins when the silkworms have completed the cocoon spinning process and entered the pupal stage.
   - Farmers delicately collect the cocoons, ensuring minimal damage to the silk threads.

2. Sorting and Grading:
   - Harvested cocoons undergo sorting based on color, size, and texture.
   - Grading is essential to separate high-quality cocoons from those of lower quality, as it influences the final silk product's characteristics.

3. Boiling the Cocoons:
   - The harvested cocoons are typically subjected to a boiling process.
   - Boiling softens the sericin, a natural protein binding the silk threads, making it easier to unravel the threads from the cocoon.

4. Unraveling the Silk Threads:
   - After boiling, skilled workers carefully unravel the silk threads from the softened cocoon.
   - The unraveling process requires precision to avoid breaking the delicate silk fibers.

5. Silk Reeling:
   - Unraveled silk threads are wound onto spools through a process known as silk reeling.
   - This step helps create uniform and continuous silk threads suitable for weaving.

6. Dyeing and Weaving:
   - Silk threads may undergo dyeing to achieve desired colors before the weaving process.
   - Weaving involves the interlacing of silk threads to create fabrics with varying textures and patterns.

7. Finishing Touches:
   - The woven silk fabric undergoes additional processes such as washing, pressing, and sometimes coating to enhance its sheen and texture.

8. Quality Control:
   - Throughout the processing stages, stringent quality control measures are implemented to ensure the final silk product meets industry standards.
   - Quality is assessed based on factors such as color fastness, strength, and overall appearance.

9. Final Products:
   - The processed silk can be transformed into a myriad of products, including garments, accessories, and home textiles.
   - The versatility of silk allows for the creation of diverse and luxurious items appreciated worldwide.

Harvesting and processing in silk farming require a harmonious blend of tradition and technology, where ancient techniques meet modern innovations. The meticulous care and attention given to each step contribute to the creation of silk products celebrated for their elegance, luster, and timeless appeal.

8. Silk Reeling and Spinning

Silk reeling and spinning are critical processes in silk farming, representing the stages where raw silk threads are carefully extracted, processed, and prepared for weaving. These intricate steps require precision and expertise to transform delicate silk fibers from cocoons into the fine threads that characterize luxurious silk textiles. Here's a detailed exploration of silk reeling and spinning in the context of silk farming:

1. Boiling and Softening:
   - The harvested cocoons are subjected to a boiling process to soften the sericin, a natural protein coating the silk threads.
   - Boiling also helps in eliminating impurities and facilitating the separation of individual silk threads.

2. Unraveling Silk Threads:
   - Skilled workers, often referred to as silk reelers, carefully unravel the softened silk threads from the cocoon.
   - This delicate process requires precision to maintain the integrity of the silk fibers and avoid breakage.

3. Silk Reeling:
   - Unraveled silk threads are wound onto a spinning wheel or a spindle in a process known as silk reeling.

- The silk reeling machine aids in achieving uniformity in the thickness and length of the silk threads.
- This step transforms the raw silk fibers into continuous and manageable strands suitable for subsequent processing.

4. Twisting and Doubling:
- After silk reeling, the silk threads may undergo twisting to enhance their strength and durability.
- Doubling involves combining multiple strands to create thicker threads, providing versatility in the types of silk yarn produced.

5. Quality Control:
- Throughout the reeling and spinning stages, strict quality control measures are implemented.
- Quality is assessed based on factors such as thread thickness, strength, and uniformity.

6. Silk Spinning:
- The spun silk threads are then ready for the weaving process.
- Traditional spinning wheels or modern spinning machines may be used, depending on the scale of production and desired thread specifications.

7. Dyeing:
- Before or after spinning, silk threads may undergo dyeing to achieve the desired color.

- Dyeing is a crucial step that adds an extra layer of artistry to the silk, allowing for a wide range of vibrant and subtle hues.

8. Weaving:
   - The spun and dyed silk threads are woven into fabrics using various weaving techniques.
   - Weaving patterns and methods contribute to the unique textures and designs of silk textiles.

Silk reeling and spinning require a delicate balance between preserving the natural qualities of silk and incorporating modern technology for efficiency. The expertise of silk farmers, reelers, and spinners plays a vital role in maintaining the exceptional quality and allure of silk textiles, ensuring that each piece reflects the rich heritage and craftsmanship of silk farming.

9. Types of Silk

Silk farming yields various types of silk, each distinguished by its origin, production method, and unique characteristics. The diverse types of silk contribute to the rich tapestry of textiles available globally. Here's a detailed exploration of some prominent types of silk:

1. Mulberry Silk (Bombyx mori):
   - The most widely produced silk, cultivated from the silkworm Bombyx mori.
   - Known for its fine texture, natural sheen, and luster.
   - Mulberry silk is versatile and used in a wide range of applications, from apparel to home textiles.

2. Tussar Silk (Antheraea spp.):
   - Produced by wild silkworms belonging to the Antheraea genus.
   - Tussar silk is characterized by its textured surface, natural gold hues, and a rustic appeal.
   - Commonly used for traditional Indian garments and decorative items.

3. Eri Silk (Samia cynthia ricini):
   - Also known as Endi or Errandi silk, produced by the silkworm Samia cynthia ricini.
   - Eri silk is unique for its short fibers, giving it a wool-like texture.

- It is often used for warm fabrics, such as shawls and blankets.

4. Muga Silk (Antheraea assamensis):
   - Originating from Assam, India, Muga silk is produced by the silkworm Antheraea assamensis.
   - Recognized for its golden-yellow color and natural sheen that improves with age.
   - Muga silk is commonly used in traditional Assamese garments.

5. Spider Silk (Various Species):
   - Though not extensively farmed due to challenges in mass production, spider silk is known for its extraordinary strength and elasticity.
   - Researchers are exploring ways to produce spider silk through genetic engineering for industrial applications.

6. Ahimsa Silk (Non-Violent Silk):
   - Produced by allowing the silkworm to complete its life cycle and emerge as a moth before harvesting the cocoon.
   - Aimed at minimizing harm to the silkworms, Ahimsa silk offers an ethical alternative to traditional silk production.

7. Bamboo Silk (Rayon or Viscose):
   - Derived from bamboo pulp, this silk alternative is known for its eco-friendly properties.

- While not produced by silkworms, it is often included in discussions of alternative silk fibers due to its silk-like characteristics.

8. Synthetic Silks (Polyester and Nylon):
   - These are man-made fibers designed to mimic the properties of natural silk.
   - While not derived from silkworms, synthetic silks offer cost-effective alternatives for various applications.

Understanding the distinct qualities of each type of silk allows consumers and artisans to make informed choices based on their preferences, ethical considerations, and the intended use of the final textile product. Silk farming continues to evolve, introducing new varieties and sustainable practices to meet the diverse demands of the global market.

10. Economic Significance

The economic significance of silk farming extends beyond its cultural and historical importance, playing a vital role in the economies of many regions worldwide. Silk production, with its intricate processes from sericulture to weaving, contributes to various sectors and supports livelihoods in both rural and urban communities. Here's a detailed exploration of the economic significance of silk farming:

1. Employment Opportunities:
   - Silk farming provides employment opportunities at various stages of the production process, from mulberry cultivation and silkworm rearing to silk processing and weaving.
   - Rural communities often rely on silk farming as a source of income and employment, contributing to poverty alleviation and economic development.

2. Cottage Industries:
   - Silk farming fosters cottage industries, especially in regions with a rich tradition of sericulture.
   - Small-scale silk production units, often family-run, contribute to the preservation of traditional craftsmanship and provide economic sustenance to local communities.

3. Global Trade and Exports:

- Silk is a highly sought-after luxury product with a global market.
- Countries with a strong tradition of silk farming, such as China, India, and Italy, are major exporters of silk products, contributing significantly to their balance of trade.

4. Textile Industry:
  - The silk textile industry, including the production of garments, accessories, and home textiles, represents a substantial economic sector.
  - Silk's unique qualities, such as its natural sheen, soft texture, and breathability, contribute to its popularity in high-end fashion and luxury markets.

5. Tourism and Cultural Heritage:
  - Regions with a strong silk farming tradition often attract tourists interested in exploring the cultural and historical aspects of sericulture.
  - Silk festivals, museums, and heritage sites related to silk production contribute to local tourism economies.

6. Innovation and Research:
  - The silk industry drives innovation and research, leading to advancements in sericulture practices, silk processing technologies, and the development of alternative silk fibers.
  - Research in genetic engineering and sustainable practices aims to enhance silk production efficiency and reduce environmental impact.

7. Diversification of Rural Incomes:
   - Silk farming provides an opportunity for diversifying rural incomes, offering an alternative or supplementary source of livelihood to agriculture.
   - Farmers can engage in mulberry cultivation and sericulture alongside traditional crops, increasing income resilience.

8. Sustainable Practices:
   - The emphasis on sustainable silk farming practices, such as organic sericulture and ethical silk production (Ahimsa silk), aligns with the growing global demand for environmentally conscious products.
   - Sustainable silk farming practices contribute to market differentiation and appeal to eco-conscious consumers.

The economic significance of silk farming reflects its multifaceted impact on local and global economies. From supporting traditional craftsmanship to driving innovation and research, silk farming remains a dynamic and culturally rich industry with far-reaching economic implications.

11. Challenges and Innovations

Silk farming, despite its rich history and economic significance, faces several challenges that necessitate continuous innovation. These challenges span from environmental and economic factors to ethical considerations. Here's a detailed exploration of the challenges and innovations in the realm of silk farming:

1. Disease and Pest Management:
   - Silkworms are susceptible to various diseases and pests, impacting cocoon quality and silk yield.
   - Innovations in disease-resistant silkworm varieties and sustainable pest management practices are crucial to maintaining healthy silkworm populations.

2. Climate Change Impact:
   - Climate change affects mulberry cultivation and silkworm rearing by altering temperature and precipitation patterns.
   - Innovations include the development of climate-resilient mulberry varieties and adapting sericulture practices to changing environmental conditions.

3.*High Labor Intensity:
   - Traditional silk farming practices often involve labor-intensive tasks, from mulberry leaf harvesting to silk reeling.

- Automation and technology-driven solutions, such as automated reeling machines, aim to reduce labor requirements and enhance efficiency.

4. Ethical Concerns:
   - The ethical considerations of sericulture, particularly the boiling of cocoons before pupation, have led to the development of ethical silk production methods, such as Ahimsa silk.
   - Ahimsa silk allows silkworms to complete their life cycle, addressing ethical concerns and catering to a conscientious consumer base.

5. Competition from Synthetic Fibers:
   - The rise of synthetic fibers poses a challenge to natural silk, as they often offer cost-effective alternatives.
   - Innovations in silk production include emphasizing its unique qualities, such as breathability and sustainability, to differentiate it from synthetic alternatives.

6. Global Market Dynamics:
   - Silk faces competition from other luxury fabrics, and market fluctuations can impact the economic viability of silk farming.
   - Diversification of silk products, marketing strategies, and exploring niche markets contribute to overcoming market challenges.

7. Sustainable Practices:

- Sustainable sericulture practices aim to minimize the environmental impact of silk farming.
- Innovations include organic sericulture methods, reducing water usage, and incorporating eco-friendly dyeing processes.

8. Research in Genetic Engineering:
- Genetic engineering research focuses on developing silkworm varieties with enhanced silk production and desirable traits.
- Genetic modifications aim to improve silk quality, disease resistance, and the overall efficiency of silk farming.

9. Promotion of Sericulture Education:
- The lack of skilled sericulturists is a challenge in some regions.
- Initiatives promoting sericulture education and training contribute to developing a skilled workforce and fostering innovation within the industry.

10. Economic Viability for Farmers:
- Ensuring economic viability for silk farmers is essential to sustain the industry.
- Innovations include fair trade practices, value addition in silk products, and supporting farmers with financial and technical resources.

The challenges faced by silk farming necessitate a dynamic approach that combines traditional knowledge with modern innovations. The industry's resilience lies in its ability to adapt to changing circumstances, embrace sustainable practices, and leverage technological advancements to secure its future in a competitive and evolving market.

12. Social aspect

The social aspect of silk farming encompasses a wide range of dimensions, including community engagement, cultural heritage preservation, and the impact of sericulture on local societies. Silk farming often plays a significant role in shaping social structures and fostering community cohesion. Here's a detailed exploration of the social aspects associated with silk farming:

1. Community Empowerment:
   - Silk farming, particularly in rural areas, becomes a source of livelihood for communities.
   - Engagement in sericulture empowers local communities by providing employment opportunities and supplementing household incomes.

2. Preservation of Cultural Heritage:
   - Regions with a strong tradition of silk farming often have deep cultural ties to sericulture.
   - The practice of sericulture, passed down through generations, becomes an integral part of the cultural heritage, fostering a sense of identity and continuity.

3. Social Cohesion and Collaboration:
   - Silk farming involves collaborative efforts, from mulberry cultivation to silk processing.

- Community members often work together, reinforcing social bonds and fostering a sense of collective responsibility.

4. Women's Empowerment:
   - In many silk-producing regions, women play a crucial role in sericulture activities.
   - Sericulture provides women with opportunities for income generation and skill development, contributing to their empowerment.

5. Cultural Celebrations and Festivals:
   - Silk festivals and celebrations are common in regions with a strong silk farming tradition.
   - These events serve as platforms for community members to come together, celebrate their heritage, and showcase their craftsmanship.

6. Education and Training:
   - Educational initiatives related to sericulture contribute to the development of a skilled workforce.
   - Training programs and workshops empower individuals with the knowledge and skills necessary for successful silk farming.

7. Social Integration in Urban Centers:
   - Urban areas engaged in silk processing and weaving create social and economic opportunities.
   - Migration of skilled workers from rural to urban centers leads to cultural exchange and contributes to the social fabric of these regions.

8. Promotion of Fair Trade Practices:
   - Fair trade initiatives aim to ensure that silk farmers receive fair compensation for their products.
   - Ethical and fair trade practices in silk farming contribute to social justice and equitable distribution of benefits.

9. Health and Well-being:
   - Sericulture practices, such as spending time in mulberry orchards, may have positive effects on the physical and mental well-being of individuals.
   - Connection with nature and traditional practices can contribute to a sense of overall health.

10. Social Entrepreneurship:
   - Silk farming can serve as a platform for social entrepreneurship initiatives.
   - Initiatives that combine sericulture with community development projects contribute to holistic social and economic development.

Silk farming, deeply intertwined with the social fabric of communities, extends its impact beyond economic considerations. The social dynamics surrounding sericulture contribute to community resilience, cultural vibrancy, and the empowerment of individuals, highlighting the intricate connection between silk farming and social well-being.

13. Environmental Dimension

The environmental dimension of silk farming encompasses a range of considerations, from the ecological impact of sericulture practices to efforts in promoting sustainability and eco-friendly methods. While silk production can have environmental implications, innovations and conscious practices aim to minimize the industry's ecological footprint. Here's a detailed exploration of the environmental aspects associated with silk farming:

1. Mulberry Cultivation and Land Use:
   - Large-scale mulberry cultivation for silkworm feeding can lead to land use changes and potential deforestation.
   - Sustainable practices focus on responsible land management, agroforestry, and organic cultivation to mitigate environmental impact.

2. Water Consumption:
   - Sericulture requires significant water usage, particularly in mulberry cultivation and silk reeling processes.
   - Water conservation efforts and the adoption of water-efficient practices contribute to minimizing the environmental footprint of silk farming.

3. Pesticide and Chemical Use:

- Conventional silk farming may involve the use of pesticides and chemicals for mulberry cultivation and pest management.
- Organic sericulture practices aim to reduce reliance on synthetic chemicals, promoting the use of natural alternatives and integrated pest management.

4. Waste Management:
- Silk processing generates waste, including discarded cocoon shells and by-products.
- Sustainable practices involve effective waste management strategies, such as recycling cocoon waste for agricultural purposes or creating eco-friendly products.

5. Energy Consumption:
- The energy-intensive nature of silk reeling processes can contribute to environmental concerns.
- Innovations in energy-efficient machinery and the use of renewable energy sources help reduce the overall energy footprint of silk production.

6. Climate Change Impact:
- Climate change can affect mulberry cultivation, silkworm rearing, and silk production processes.
- Adaptation strategies and the development of climate-resilient sericulture practices contribute to mitigating the impact of changing climatic conditions.

7. Biodiversity Conservation:
   - Large-scale monoculture of mulberry plants may have implications for local biodiversity.
   - Sustainable sericulture practices incorporate agroecological approaches that support biodiversity, such as maintaining diverse cropping systems and preserving natural habitats.

8. Genetic Engineering and Innovation:
   - Research in genetic engineering aims to develop silkworm varieties with enhanced silk production and resilience to environmental stress.
   - Innovations in silk farming technologies contribute to sustainable practices and reduce the environmental impact of conventional sericulture.

9. Eco-Friendly Dyeing Processes:
   - Silk dyeing processes can involve harmful chemicals that impact the environment.
   - Eco-friendly dyeing methods, such as natural dye extraction from plants, contribute to reducing the environmental impact of silk processing.

10. Certification and Standards:
   - Certification programs and standards, such as organic and eco-friendly certifications, help consumers identify silk products produced with environmentally responsible practices.
   - Adoption of these certifications encourages producers to adhere to sustainable and eco-friendly methods.

While silk farming poses environmental challenges, ongoing efforts in research, innovation, and the adoption of sustainable practices contribute to minimizing its impact on ecosystems. The intersection of environmental considerations with silk farming reflects a broader commitment to balance economic activities with ecological sustainability.

14. Competition

Competition is a significant factor in the silk farming industry, influencing market dynamics, pricing, and the overall sustainability of silk production. The competition within the silk sector involves various aspects, from the global market for silk products to the challenges posed by alternative fibers. Here's a detailed exploration of the competition in the context of silk farming:

1. Global Market Dynamics:
   - Countries with a long tradition of silk farming, such as China, India, and Italy, dominate the global silk market.
   - Intense competition exists among these key players, influencing silk prices and market trends.

2. Alternative Luxury Fibers:
   - Silk faces competition from other luxury fibers like cashmere, wool, and specialty textiles.
   - The choice of consumers between different high-end fibers can impact the demand for silk products.

3. Synthetic Fiber Alternatives:
   - The rise of synthetic fibers, including polyester and nylon, poses a challenge to natural silk.
   - Synthetic alternatives offer cost-effective solutions, and their popularity impacts the market share of natural fibers like silk.

4. Market Differentiation:
  - Silk farmers and producers often engage in market differentiation strategies to distinguish their products.
  - Emphasizing the unique qualities of silk, such as its natural sheen, breathability, and luxurious feel, helps differentiate it from competing fibers.

5. Niche Markets and Specialized Products:
  - Producers explore niche markets by offering specialized silk products, such as organic silk, Ahimsa silk, or handwoven silk.
  - Targeting specific consumer preferences contributes to carving out unique market segments.

6. Innovation in Product Development:
  - Continuous innovation in silk product development, including diverse weaves, patterns, and applications, helps maintain competitiveness.
  - Unique and novel silk products can attract consumers looking for distinctive and high-quality items.

7. Sustainable Practices:
  - The growing demand for sustainable and eco-friendly products has led to an emphasis on sustainable silk farming practices.
  - Producers adopting environmentally conscious methods may gain a competitive edge in markets where sustainability is a significant factor.

8. Marketing and Branding:

- Effective marketing strategies and branding efforts play a crucial role in positioning silk products in the market.
- Establishing a strong brand image can influence consumer perceptions and drive demand.

9. Economic and Political Factors:
- Economic fluctuations and geopolitical factors can impact the silk industry, affecting production costs and market accessibility.
- Unforeseen economic challenges may intensify competition or create opportunities for certain producers.

10. Supply Chain Efficiency:
- The efficiency of the silk supply chain, from mulberry cultivation to the final product, influences cost competitiveness.
- Producers investing in streamlined and efficient supply chain processes may gain advantages in terms of cost and quality.

Understanding and navigating competition are crucial for silk farmers and producers to thrive in a dynamic market. Adapting to changing consumer preferences, embracing sustainable practices, and continuous innovation are key strategies to stay competitive and ensure the long-term viability of silk farming.

15. Improving Production

Improving production methods in silk farming is essential for enhancing efficiency, sustainability, and the overall quality of silk products. Innovations in sericulture practices contribute to addressing challenges, reducing environmental impact, and meeting the demands of a dynamic market. Here's a detailed exploration of ways to improve production methods in silk farming:

1. Mulberry Cultivation Techniques:
   - Implementing advanced cultivation techniques, such as precision farming, can optimize mulberry yield.
   - Introducing disease-resistant mulberry varieties enhances productivity and reduces the need for chemical interventions.

2. Climate-Resilient Sericulture:
   - Developing sericulture practices resilient to climate change involves adapting to shifting weather patterns.
   - Research focuses on creating silkworm varieties and mulberry cultivars that can thrive in varying climatic conditions.

3. Integrated Pest Management (IPM):
   - Adopting IPM strategies reduces dependence on chemical pesticides.

- Encouraging natural predators, using biopesticides, and implementing crop rotation are integral to sustainable pest management.

4. Automated Silkworm Rearing:
   - Automation in silkworm rearing processes, such as feeding and cocoon harvesting, reduces labor intensity and improves efficiency.
   - Robotic systems and sensor technologies can contribute to precise monitoring and management of silkworm colonies.

5. Energy-Efficient Silk Reeling Machines:
   - Upgrading silk reeling machines to be more energy-efficient helps reduce the overall energy consumption in the silk production process.
   - Innovations in machinery design contribute to improved silk reeling without compromising on quality.

6. Genetic Engineering for Silkworms:
   - Research in genetic engineering aims to develop silkworm varieties with enhanced silk production and resistance to diseases.
   - Genetic modifications can lead to silkworms with improved silk quality, faster growth rates, and reduced susceptibility to environmental stress.

7. Water-Conserving Practices:
   - Implementing water-conserving irrigation techniques in mulberry cultivation minimizes water usage.

- Adoption of drip irrigation systems and rainwater harvesting contribute to sustainable water management.

8. Organic Sericulture Practices:
   - Transitioning to organic sericulture practices reduces reliance on synthetic chemicals and promotes environmental sustainability.
   - Organic sericulture involves using natural fertilizers, pest control through biological means, and eco-friendly disease management.

9. Value Addition and Diversification:
   - Exploring value addition in silk products through innovative weaving patterns, designs, and blends with other fibers enhances market appeal.
   - Diversifying silk product offerings caters to diverse consumer preferences and expands market opportunities.

10. Education and Training Programs:
    - Promoting education and training programs in sericulture ensures a skilled workforce.
    - Training initiatives equip farmers and workers with the latest knowledge in advanced sericulture practices, contributing to improved production methods.

11. Digital Technologies for Monitoring:
    - Utilizing digital technologies, such as IoT sensors and data analytics, enables real-time

monitoring of environmental conditions in silkworm rearing facilities.
  - Digital solutions enhance precision in managing factors like temperature, humidity, and ventilation, optimizing conditions for silkworms.

12. Quality Control Measures:
  - Implementing stringent quality control measures throughout the production process ensures the consistency and high quality of silk products.
  - Automated inspection systems and standardized protocols contribute to maintaining product standards.

Improving production methods in silk farming requires a multidimensional approach, integrating technological advancements, sustainable practices, and continuous research and development. These efforts not only contribute to the efficiency and competitiveness of the silk industry but also address environmental and social considerations, ensuring a holistic and sustainable approach to silk farming.

16. Silk Products

Silk farming is inherently tied to the diverse range of exquisite silk products that captivate consumers globally. The culmination of a meticulous process, from sericulture to weaving, results in a myriad of luxurious items known for their sheen, softness, and timeless elegance. Here's a detailed exploration of various silk products that showcase the versatility and allure of silk:

1. Silk Apparel:
   - Silk is widely celebrated in the fashion industry, giving rise to a vast array of garments.
   - Dresses, blouses, shirts, skirts, and ties crafted from silk exhibit a natural luster, drapability, and comfort, making them ideal for both casual and formal wear.

2. Scarves and Shawls:
   - Silk scarves and shawls are coveted accessories, prized for their lightweight feel and ability to add a touch of sophistication to any outfit.
   - The natural breathability of silk makes it a popular choice for scarves that can be worn throughout the year.

3. Lingerie and Sleepwear:
   - The smooth, soft texture of silk lends itself to intimate apparel.
   - Silk lingerie and sleepwear offer a luxurious and comfortable experience, with its natural

temperature-regulating properties contributing to a blissful night's sleep.

4. Silk Ties and Pocket Squares:
   - Silk's refined appearance makes it an ideal material for accessories like ties and pocket squares.
   - The sheen and texture of silk ties add a polished finish to formal attire, while pocket squares provide a subtle touch of elegance.

5. Bedding and Linens:
   - Silk beddings, including sheets, pillowcases, and duvet covers, are known for their smoothness and ability to promote a comfortable sleep environment.
   - Silk's hypoallergenic properties make it an attractive choice for those with sensitive skin.

6. Home Textiles:
   - Silk extends its luxury to various home textiles, including curtains, drapes, and upholstery fabrics.
   - The rich colors and sheen of silk enhance the aesthetic appeal of living spaces, creating an atmosphere of opulence.

7. Traditional Ethnic Garments:
   - In many cultures, silk holds cultural significance and is used to create traditional garments.
   - Examples include sarees in India, kimono in Japan, and ao dai in Vietnam, showcasing the diverse cultural applications of silk.

8. Silk Accessories:
   - Silk is incorporated into various accessories, such as handbags, gloves, and hats.
   - The fine texture and durability of silk enhance the overall quality and appeal of these accessories.

9. Wedding Dresses:
   - Bridal couture often features silk wedding dresses, celebrated for their elegance and timeless beauty.
   - Silk's ability to drape gracefully contributes to the creation of stunning bridal gowns.

10. Silk Art and Craft:
    - Silk serves as a canvas for artistic expression, with hand-painted silk scarves, wall hangings, and art pieces showcasing the craftsmanship of artisans.
    - The absorbent nature of silk allows for vibrant and intricate designs.

11. Silk Upholstery for Furniture:
    - Silk upholstery fabrics are used to add a touch of luxury to furniture.
    - The sheen and smooth texture of silk contribute to creating sophisticated and inviting living spaces.

12. Specialty Silks:
    - Specialized silk products include variations like Ahimsa silk (non-violent silk), where the silkworm is

allowed to complete its life cycle, and Eri silk, known for its wool-like texture.

   - These specialty silks cater to specific consumer preferences, emphasizing ethical considerations and unique textures.

Silk products, ranging from everyday wearables to opulent furnishings, exemplify the timeless allure of this luxurious fabric. The craftsmanship involved in silk farming and processing is reflected in the diversity of silk items that continue to capture the imagination of individuals worldwide.

17. Uses of Silk Beyond Textiles

Silk is primarily composed of fibroin, a protein, and sericin, a gum-like substance. Fibroin forms the structural basis of silk, providing strength and elasticity, while sericin acts as a bonding agent. The chemical composition of silk includes amino acids, such as glycine, alanine, and serine, which contribute to its unique properties.

Aside from its use in textile production, silk has various applications:

1. Biomedical:
   -Surgical sutures and implants: Silk's biocompatibility, strength, and controlled degradation make it ideal for sutures and implants that are eventually absorbed by the body.
   -Drug delivery systems: Silk can be used to encapsulate and release drugs in a controlled manner, providing targeted drug delivery and reducing side effects.
   -Tissue engineering: Scaffolds made of silk can be used to support the growth and regeneration of tissues, such as skin, bone, and cartilage.

2. Biotechnology:
   Enzymes and proteins: Silk can be used to immobilize enzymes and proteins, improving their stability and reusability in biocatalytic processes.

-Biosensors: Silk can be used to create sensitive biosensors for detecting various biological molecules, such as glucose and antibodies.

-Biomaterials: Silk can be combined with other materials to create biocompatible and biodegradable composites for various applications.

3. Biodegradable Packaging:
   - Silk proteins can be processed into biodegradable films, offering a sustainable alternative to traditional packaging materials. These films have low environmental impact.

4. Cosmetics and Skincare:
   - Sericin, extracted from silk, is used in cosmetics and skincare products due to its moisturizing and antioxidant properties. It helps improve skin texture and elasticity.

5. Optics and Electronics:
   - Silk fibers can be used in optical and electronic devices. Researchers are exploring silk's potential for creating biodegradable optical components and flexible electronics.

6. Wound Healing:
   - Silk's biocompatibility extends to wound healing. Silk-based dressings provide a conducive environment for healing, and silk fibroin has been investigated for its role in promoting tissue regeneration.

7. Insecticides and Coatings:
   - Silk can be utilized as a carrier for delivering insecticides in agriculture. Additionally, silk coatings on fruits can extend their shelf life by reducing moisture loss and preventing microbial growth.

8. Bulletproof Clothing:
   - Researchers have explored using silk fibers in the development of lightweight, flexible, and durable bulletproof clothing. The unique combination of strength and flexibility in silk makes it a potential candidate for protective gear.

9. Musical Instruments:
   - Some traditional musical instruments, like the strings on certain types of harps, are made from silk due to its unique sound-producing qualities.

10. Food and Fertilizer:
   - Silk, primarily known for its luxurious texture and use in textiles, is produced by silkworms. Interestingly, silkworms are not only valuable for silk but also serve as a by-product with diverse applications.
   - Silkworms are consumed in certain cultures as a delicacy. In parts of Asia, especially China and Thailand, boiled or fried silkworm pupae are enjoyed as a crunchy and protein-rich snack. The pupae are a good source of essential amino acids, healthy fats, and minerals, making them a unique dietary option.

- Silkworms are nutritionally dense, containing proteins, fats, vitamins, and minerals. They are particularly rich in protein, making them a potential source of essential amino acids crucial for human health. Additionally, silkworms provide unsaturated fats and micronutrients, contributing to a well-rounded nutritional profile.

- Silkworm excrement, also known as silkworm castings or frass, serves as a valuable organic fertilizer. Rich in nutrients like nitrogen, potassium, and phosphorus, silkworm castings enhance soil fertility and promote plant growth. Farmers often use this by-product to improve soil quality and boost crop yields in an eco-friendly manner.

This list demonstrates the diverse potential of silk beyond its traditional textile uses. As research continues, we can expect to see even more innovative applications for this remarkable natural material.

Conclusion

As we reach the final threads of this exploration into the intricate world of silk farming, I hope you've found the journey as enlightening as it has been for me. From the humble silkworm to the global tapestry of silk trade, "Unraveling the Secrets of Silk Farming" seeks to leave you with a newfound appreciation for the delicate interplay between nature, culture, and commerce. As we bid farewell to this silken odyssey, may it linger in your thoughts, much like the enduring allure of silk itself.

Sincerely,
 M.I.Fazil